D1497748

Farming

Sue Hadden

Our Green World

Acid Rain
Atmosphere
Deserts
Farming
Oceans
Polar Regions
Rainforests
Recycling
Wildlife

Cover: Combine harvesters cut the crop in an enormous
wheat field in the USA.

Book editor: Anna Girling
Series editor: Philippa Smith
Series designer: Malcolm Walker

First published in 1991 by
Wayland (Publishers) Ltd
61 Western Road, Hove
East Sussex BN3 1JD, England

British Library Cataloguing in Publication Data
Hadden, Sue
   Farming. – (Our green world)
   I. Title II. Series ✓
 ⌣ 630

   ISBN 0-7502-0326-9

Typeset by Kudos Editorial and Design Services, Sussex, England
Printed in Italy by G. Canale & C.S.p.A., Turin
Bound in France by AGM

Contents

Words printed in **bold** in the main text are explained in the glossary on page 45.

How farming began

Think of the countryside – trees, flowers, farms and fields. But thousands of years ago there were no fields or farms. Instead, most of the land was covered in thick forests. People lived by hunting wild animals for food and eating wild plants.

▼ *In parts of the USA wheat fields cover huge areas of land. Long ago, these areas were wild grasslands.*

▲ *Bread wheat (left) has larger grains than other grasses.*

▲ *A farmer with some young wheat plants.*

About 12,000 years ago people learned to tame animals such as goats and sheep. They kept them in small herds and moved around to find wild plants to feed them.

Then, about 9,000 years ago, people living in the Middle East discovered wild wheat. It had large grains that were good for making flour for bread. The people needed more of this wheat so they started sowing grains in the soil. And so they became the first crop farmers.

▲ *People in Peru grow maize high up in the mountains.*

In different parts of the world farmers grow different crops. Flat parts of Europe and North America are good for growing wheat and barley. In Asia farmers grow rice. High up in the mountains of South America people grow maize. All these crops are cereals.

A few old beech woods still grow in Europe. ▶

All over the world farming has changed the **environment**. To make open fields for crops, people have cut down forests and drained water from wet, marshy land. Until about 5,000 years ago most of Europe was covered in forests. Today most parts of Europe have more fields than forests.

◀ *These hills in southern England were covered in trees before early farmers cleared them for sheep to graze.*

7

When a forest is cut down to grow crops, a lot of the wildlife disappears too. But fields can make a good home for some animals and plants. Mice, birds and insects like to eat grain from the crops. Many kinds of grasses and wild flowers can grow in hay meadows.

Heathlands are now found in places where early farmers burned down forests to grow crops. Heathlands are a good wildlife **habitat**.

▲ *Heather and gorse are lovely heathland plants.*

Sadly, even these newer habitats are disappearing. As farmers try to grow more and more food on the land, there is less room for wildlife.

Hedges

Hedges are very important for wildlife. Often they are hundreds of years old and are made from many kinds of trees and shrubs.

A hedge is a good shelter for small animals, such as mice. The thick twigs make a safe nesting place for small birds. And many lovely wild flowers grow in the shelter at the bottom of a hedge.

▼ *A woodmouse makes its home in a hedge.*

Bottom left
Honeysuckle and foxgloves grow in this hedge.
Bottom right
Most farmers use machines to cut the hedges. ▼

Today farmers put **fertilizers** containing **chemicals** on their fields, to make their crops grow better. But often the crops do not take up all the fertilizer. Some goes into the ground or is washed away by rain. Then the chemicals can get into streams, rivers and ponds, where they cause **pollution**.

◀ *A farmer spreads fertilizer on to a field.*

Polluted ponds and lakes

Fertilizers contain chemicals called nitrates and phosphates. These help plants on land to grow. But if they get into ponds and lakes, they also make water plants called **algae** grow very fast. Soon the algae cover the lake surface in a thick green slime. Then tiny **organisms** called **bacteria** feed on the algae. The bacteria spread quickly and use up all the oxygen in the water. This kills fish and pond snails which need oxygen to live.

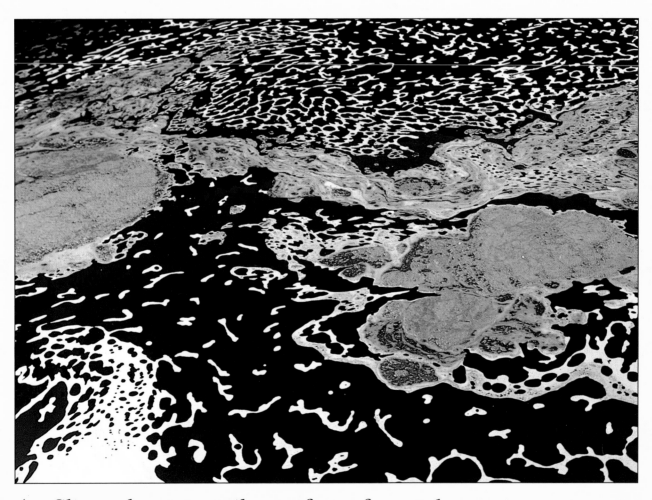

▲ *Slimy algae cover the surface of a pond.*

▲ *A farmer piles up liquid slurry in a pit.*

Farmers used to put manure on fields to help crop plants grow. Manure is made from farm animal dung mixed with straw. But today most farmers use a very strong, liquid manure called slurry.

Slurry is made from the dung and **urine** produced by cattle and pigs. Large amounts of slurry have to be stored in pits on the farm. It smells unpleasant and, if it leaks out of the pit, it can run into ponds and streams, causing pollution.

Farmers must store plenty of grass to feed to their cattle in winter. Instead of cutting grass to make hay, many farmers now use silage. This is made from grass cuttings that have been packed down and left to **ferment**.

Silage can also be used as a kind of manure to help crops grow. But often the silage grass is too wet and makes a sour-smelling liquid. If the liquid silage runs off the field, it can pollute rivers and lakes.

▼ *A farmer uses a tractor to squash down grass cuttings for silage.*

Life is not easy for farmers. They work hard to make their crops grow and then plenty of insects and other animals come and eat them! We call them pests. Farmers kill pests by spraying the crops with poisonous chemicals called **pesticides**.

Farmers do not want weeds spoiling their crops, either. So they use poisons called **herbicides** to kill them.

▼ *You could not eat this maize. It is covered with a black mould.*

▲ *A swarm of locusts in northern Africa.*

Some insect pests completely ruin a good crop. Swarms of locusts eat up huge areas of crops in a few hours.

▲ *Crop sprays can harm wildlife and people, too. This man is not wearing a mask to protect him.*

Pesticides kill insect pests such as aphids (greenfly) and grain weevils, but they often poison useful insects too, such as butterflies and ladybirds. Fieldmice and birds of prey, such as owls and falcons, can also be poisoned. Even people can become ill if they breathe in these powerful chemicals.

Mice and birds can be poisoned when they eat crops sprayed with pesticides. But if a bird of prey eats a poisoned mouse or bird, it too can be harmed. Some pesticides are now forbidden in some countries, to protect the wildlife.

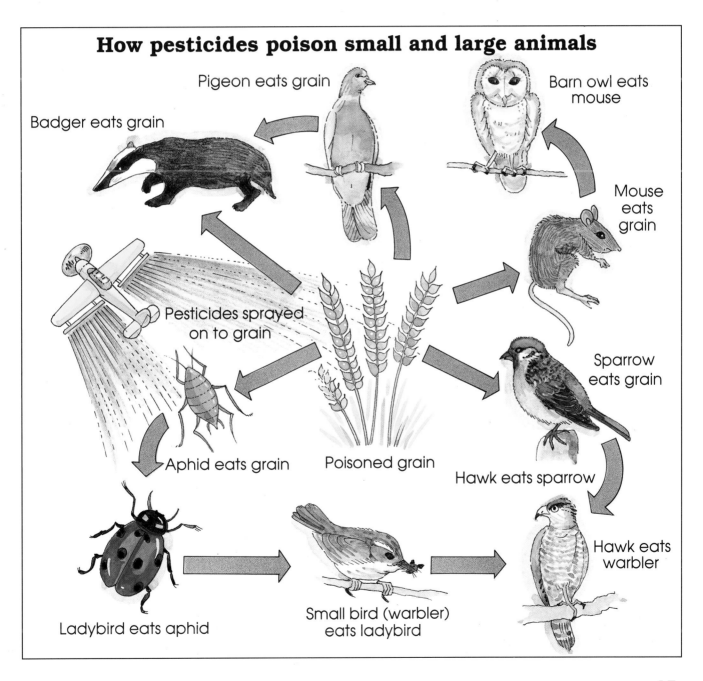

How pesticides poison small and large animals

Badger eats grain

Pigeon eats grain

Barn owl eats mouse

Mouse eats grain

Pesticides sprayed on to grain

Aphid eats grain

Poisoned grain

Sparrow eats grain

Hawk eats sparrow

Ladybird eats aphid

Small bird (warbler) eats ladybird

Hawk eats warbler

Farming and wildlife

▲ *A beautiful hay meadow in the Alps in France.*

Some old ways of farming are good for wildlife. Hay meadows are full of wild flowers and buzz with insects in summer. Farmers let the plants flower before they cut the grass to feed their cattle in winter.

The only flowers in this field belong to the crop of bright yellow rape.

Not many hay meadows are left today. Fields are often just full of grass used for silage. Chemical sprays stop wild flowers from growing among the crops. Large areas of wild grassland have been turned into crop fields. So wild flowers are less common than they used to be.

A field of grass is cut for silage.

Animals and plants are disappearing all over the world, as humans destroy their habitats. Forests, wild grasslands and marshes are often cleared so that farmers can grow more crops. Animals like wolves and bears have disappeared from many countries.

Barn owls in danger

Barn owls hunt mice and voles in the countryside. They used to be common on farms, where they nested in old barns and farm buildings.

Now that most fields are sprayed with pesticide, many mice and voles have disappeared. Also, many old barns have been pulled down. So barn owls have lost their food and their nesting places, and have become rare.

▲ *A barn owl hunts for food at night.*

Some farmers in Europe are now trying to bring barn owls back. They have stopped spraying pesticides on to their fields and have provided nesting sites for the birds.

What happened to the wolf?

Like barn owls, wolves used to be common in Europe. They lived in the big forests which covered much of the land. However, people thought that wolves could kill humans. Farmers disliked wolves because they attacked their sheep.

So people began to hunt wolves and many were killed. Also, more and more of the forests were cleared to make farmland. Wolves became **extinct** in Britain in 1550. Only a few wolves survive today, in Scandinavia and North America.

◀ *Wolves look fierce but they are shy animals which live peacefully in groups.*

Marshes and wetlands are very good places for rare wild flowers, water birds and insects such as dragonflies. However, wetlands can cover large areas of land which farmers may need for growing crops. Many wetlands have been drained to make crop fields. When this happens, the rich wildlife disappears.

◀ *Lovely wild flowers like marsh marigolds and orchids grow in damp, marshy places.*

Today, bison only live in national parks.

The North American prairies were once natural grassland, where huge herds of bison (wild cattle) used to graze. Now much of the prairies has become wheat fields and the bison have disappeared.

The western swamp turtle of Australia is now rare because its swampy home is being drained for farmland.

Forests all over the world have been cut down to make farmland. But **rainforests** are disappearing faster than any other kind. They are special because they contain more animals and plants than anywhere else on Earth.

Rainforests grow in Central and South America, Africa and South-east Asia. Huge areas have been cut down for their **timber** or to make way for farmland.

This green basilisk lizard is one of many unusual rainforest animals. ▶

◀ *Rainforests have been cut down to make grazing land for beef cattle.*

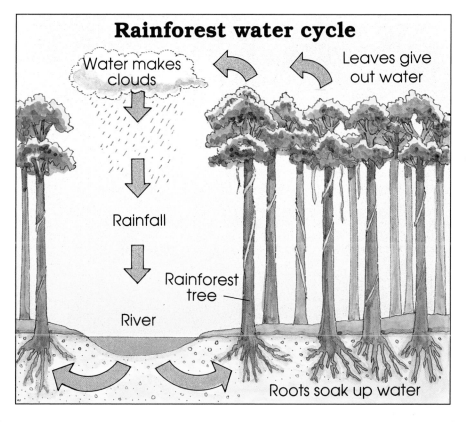

Rainforest water cycle

Water makes clouds

Leaves give out water

Rainfall

Rainforest tree

River

Roots soak up water

◄ *Rainforest trees keep the* **rainfall cycle** *going.*

What happens when rainforests are cut down? First, the many wild plants and animals disappear. Then the **climate** changes, too. Huge rainforest trees soak up rainwater with their roots. Later their leaves return the water to the air. Without the trees, the rain stops falling. The treeless area can become dry and dusty, like a **desert**.

▲ *The world's smallest monkey lives in the rainforest, but it can also live in crop fields.*

Soil and water

Plants need good soil and fresh water to grow. In some parts of the world farmers only have poor soil, and sometimes little water.

▼ *In parts of Asia, farmers grow rice on steep mountains. They make terraces to stop the soil being washed away, and to hold the water for the plants to grow in.*

▲ *These hills were once covered in rainforest. Now that the trees have gone, the soil has been washed away.*

Soil can be blown away by the wind or washed off the land by rain. This is called soil erosion and it is a big problem for farmers. Soil erosion often happens when people cut down forests on mountains. The roots of the trees kept the soil in place. Without them it is soon blown or washed away.

The soil builds up in other places, blocking rivers. When heavy rains come, the blocked-up rivers burst their banks, causing floods.

Soil erosion is also a problem in countries with **droughts**, for example in some African countries. Without rain, the crops wither and die. Animals have to graze on wild shrubs. When these disappear, there is nothing to hold the soil and it blows away.

▼ *These people in Sudan move around to find grazing for their animals. Dry areas soon become overgrazed.*

The American Dust Bowl

During the 1930s there was a long drought in part of the USA. The soil turned to dust and blew away. Crops could not grow and farm animals died. Farming families had to leave their homes.

The useless dusty soil covered the farmland for many years. The whole area was called the Dust Bowl. Now, farmers can grow crops there again. But if another long drought comes, the same thing could happen again.

▲ *An American farmer is glad to see rain after a drought in 1988.*

▼ *This old photograph shows a farm in the Dust Bowl.*

▲ *A sprinkler sprays water on to a field in Utah, USA.*

Farmers can grow crops in places with little rainfall by using stored water. Some rainwater collects underground and people can reach this stored groundwater by digging wells. Then they pass the water along pipes or channels to the fields. Watering the land in this way is called irrigation.

If farmers take too much water from under the ground, the stores will dry out. Also, if too much water is used on the land, it leaves behind a layer of salt on the soil. Then, crops cannot grow in the salt.

A layer of salt lies on this Egyptian field. ▶

An irrigation channel in Egypt. Look at the lush crops around it. ▼

Feeding the world

Since the 1950s, the world **population** has grown quickly, especially in **developing countries**. In the 1960s and 1970s the governments of many developing countries tried to find ways to feed the millions of people. They used modern farming methods to grow new kinds of wheat and maize, which were able to survive in hot countries. This was called the Green Revolution.

▼ *Cocoa plants being sprayed with expensive pesticide.*

▲ *Pineapples growing in Kenya, Africa. Most of the fruit will be sold to other countries, to earn money.*

The new methods of the Green Revolution did grow more food, but they were very expensive. The poorer countries had to buy fertilizers and pesticides from richer countries. To help pay for these expensive chemicals, the farmers sold some of the crops to other countries. So many people in the developing countries still went hungry.

◀ *A grain store in Canada. This country grows plenty of wheat.*

▼ *A market in Ethiopia. In many parts of Africa, there is not much food to go round.*

In Europe and North America, the climate is good for crops. People have money to buy fertilizers. These countries grow plenty of food – even too much.

In developing countries with dry climates, crops do not grow so well. Most farmers are poor and cannot buy fertilizers. Sometimes **famines** occur.

▲ *Farm chickens crowded together in cages. This method produces cheap meat and eggs.*

Today, farmers in developed countries produce more food than ever before. Lots of animals are kept in a small area, sometimes indoors. This saves space and makes it easier to look after the animals. However, disease spreads quickly and farmers have to give the animals medicines. Some people think this is an unnatural way to keep animals.

When chickens are kept using modern methods, there is more chance that their eggs will have **salmonella**. Salmonella can give people bad stomach upsets. Now more people like to eat eggs and meat from **free-range** chickens.

▲ *When we buy meat, do we know how it was produced?*

▼ *Many people now try to eat good, healthy food.*

▲ *Cattle eating special feed instead of grass.*

A few years ago, scientists discovered mad cow disease, or BSE. Cattle with this brain disease become very ill and cannot stand up properly. It may be passed to cattle when they are fed meat from sheep with a similar disease. This is a modern method of feeding cattle. Naturally, cows eat grass. Scientists are trying to stop the disease from spreading.

Farming in the future

Farmers are very important. Without them it would be impossible for everyone to grow enough food for themselves. Farmers now produce more food than ever before, but the modern methods that help them to do this can harm the environment. Is there a way of farming that helps people and the environment?

Crops of many kinds

Important crops like wheat and maize are grown all over the world. But there are many thousands of unknown plants which might be useful as food or medicines. Many of them grow in rainforests. Scientists are saving some of their seeds because we may need them in the future.

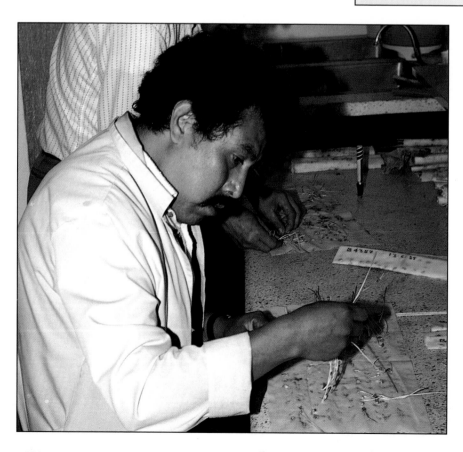

◀ *This man is testing different kinds of wheat seeds.*

Pests are one of the farmer's biggest problems. But instead of using strong chemicals to kill them, farmers can use nature's own pest control. For example, farmers and gardeners are now bringing in ladybirds to eat up insect pests such as aphids.

◀ *Ladybirds eat aphids, which are a pest of many plants.*

South American cane toads were taken to Australia to control a pest beetle. But the toads did not eat all the beetles – they spread to become a pest themselves! ▶

▲ *Poppies and camomile grow round the edge of a field in Germany.*

Farmers can help nature by letting the edges of their fields grow wild. Then wild flowers and grasses will grow there, attracting butterflies and small animals. In this way, wildlife can live in fields without spoiling the farmer's crops.

How can farmers help the environment?

Farmers can help nature if they:
- Use natural farming methods without chemical crop sprays.
- Leave some areas for wild flowers and animals to live in.
- Leave ponds and trees on their land. These are good habitats for wildlife.
- Look after hedges. These are good habitats for wildlife and help to stop soil erosion.

Farmers can use natural methods to make their crops grow strong and healthy. Instead of using chemical fertilizers, more farmers are putting natural manures on the land. These natural ways of growing food are called organic farming. This is becoming popular with farmers and the public.

This farmer has left trees on his farm. He knows that they are good for wildlife.

In the future, perhaps more farmers will go back to using natural methods. But many farmers will always want to use the latest farming ideas.

Already scientists are making new kinds of plants which can make their own chemical pest control. Not everyone thinks this is a good idea. Some people are worried that these 'human-made' plants are unnatural and could harm the environment.

▲ *Scientists are trying many experiments with plants.*

Another problem for the future is the greenhouse effect. The Earth's atmosphere seems to be getting warmer, and this could cause bad floods.

The greenhouse effect

New farming methods can damage the environment, but all of us damage the natural world in some way. Our modern way of life causes a lot of pollution. It may even be causing the temperature of the world to rise. We call this the greenhouse effect, and it works like this.

Industrial countries need energy to run power stations, factories and all the electrical equipment in people's homes. To make the energy, they burn **fossil fuels** – coal, oil and gas. When fossil fuels are burnt, they give off a gas called carbon dioxide. This gas traps some of the heat from the sun inside the Earth's atmosphere, and so the Earth is slowly getting warmer. By the year 2050 the Earth may be 4 °C warmer than today. That could be enough to melt ice in the polar regions, causing sea levels to rise and terrible floods.

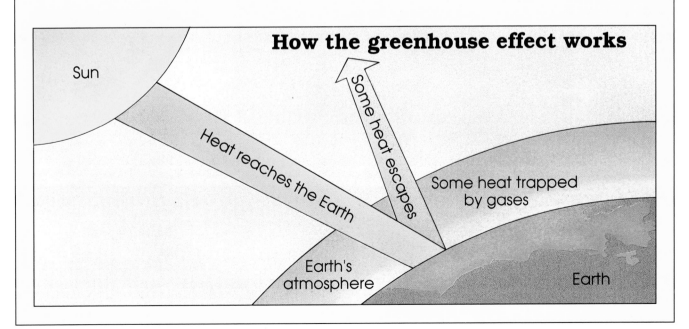

How the greenhouse effect works

Sun

Heat reaches the Earth

Some heat escapes

Some heat trapped by gases

Earth's atmosphere

Earth

Large areas of the rainforest in South America have been cleared to grow crops.

This rainforest in Africa is protected as a national park. ▶

The cutting down of rainforests is another very serious problem for the world. Farmers in many developing countries need to grow crops in places where there are rainforests. But as people understand why rainforests are so special, they are making some of them into national parks. So some rainforests will remain, while others are cleared. For the future of our planet, people and nature must live together.

Glossary

Algae Simple plants that live in water.

Bacteria (singular **bacterium**) A large group of one-celled organisms. Bacteria can break down plant and animal material.

Chemicals Substances which are used in artificial fertilizers and pesticides.

Climate The general weather conditions in a part of the world.

Desert A place where nothing grows, usually because it is too hot and dry.

Developing countries Countries in Africa, South America and Asia which do not have a developed industry.

Drought A long period of dry weather without rain.

Environment The natural world around us – for example, plants, animals, rivers and rocks.

Extinct No longer living on the Earth.

Famine A serious shortage of food which leads to many people starving.

Ferment Slowly decompose or rot away.

Fertilizer A substance put on plants to help them grow.

Fossil fuels Coal, oil and gas used to provide energy. They were formed over millions of years from living organisms.

Free-range Allowed to move freely in an open space.

Habitat A place where plants and animals live. Meadows, woods and ponds are all kinds of habitat.

Heathland An unusual habitat with heather, gorse and sometimes pine trees.

Herbicide A chemical which kills unwanted wild flowers and weeds.

Organism A living animal or plant. It often means a very small creature.

Pesticide A chemical which kills crop pests, like insects.

Pollution Anything which spoils the environment and harms wildlife, such as smoke from factory chimneys, poisonous chemicals and litter.

Population The number of people living in a town, country or the world.

Rainfall cycle The process in which rain falls to the ground and then evaporates to form clouds, ready to rain again.

Rainforest A forest with tall trees which grows in tropical parts of the world.

Salmonella A bacterium which can live in chickens and eggs.

Timber The wood from trees.

Urine The liquid that is passed out of the body.

Finding out more

Books to read

For younger readers:
Habitat Destruction by Tony Hare (Franklin Watts, 1991)
Ian and Fred's Big Green Book by Fred Pearce (Kingfisher Books, 1991)
Let's Visit a Farm series by Sarah Doughty and Diana Bentley (Wayland, 1989–90)
Rainforest Destruction by Tony Hare (Franklin Watts, 1990)

For older readers:
Food and Farming by John and Sue Becklake (Franklin Watts, 1991)

Useful addresses

Australian Association for Environmental Education
GPO Box 112
Canberra ACT 2601

Environment and Conservation Organizations of New Zealand (ECO)
P.O. Box 11057
Wellington

Farming and Wildlife Trust
National Agricultural Centre
Stoneleigh
Near Kenilworth
Warwickshire CV8 2RX

Friends of the Earth (UK)
26–28 Underwood Street
London N1 7JQ

Friends of the Earth (Canada)
Suite 53
54 Ottawa Street
Ottawa KP5 CS

Friends of the Earth (Australia)
National Liaison Office
366 Smith Street
Collingwood
Victoria 3065

Friends of the Earth (NZ)
Nagal House
Courthouse Lane
PO Box 39/065
Auckland West

Greenpeace (UK)
30–31 Islington Street
London N1 8XE

Greenpeace (Australia)
310 Angas Street
Adelaide 5000

Greenpeace (Canada)
2623 West 4th Avenue
Vancouver
BCV6K 1P8

Picture acknowledgements
The photographs in this book are by: Bruce Coleman Ltd 4 (Nicholas Devore), 7 below (Roger Wilmhurst), 9 left (Dennis Green), 9 right (N. Blake), 11 (Frieder Sauer), 18 (Hans Reinhard); Flour Advisory Bureau 5; FWAG/Helen Simonson 41; Hulton-Deutsch Collection 29 below; Hutchison Library 16 (Sarah Errington), 24 below, 27 (Taylor), 31 above, 38 (Dr. Nigel Smith), 44 below (P. Parker); Mark Lambert 13, 14, 19 below; Oxford Scientific Films 7 above and 8 (Terry Heathcote), 9 above (Colin Milkins), 19 above (Ronald Toms), 23 below (Babs and Bert Wells), 24 above (Michael Fogden), 25 (Michael Dick), 35 (Zigmund Leszczynski), 39 above (Raymond Blythe), 39 below (Kathie Atkinson), 44 above (J. Devries); Tony Stone Worldwide *cover*; Topham Picture Library 12, 29 above, 30 (John Griffin), 32, 33, 34 above and below, 42 below; ZEFA 6, 10, 15 (Tortoli), 20 (Heintges), 21, 22, 23 above, 26, 28 (Zingel), 31 below (F. Damm), 36, 37 (Kummels), 40 (Streichan). The illustrations are by Stephen Wheele.

Index

DATE DUE